はしがき

　この多目的結束バンド外し具は、一度、締めて、結束すると、再利用ができないタイプの結束バンドを緩めて、再利用をできるようにしたものである。これにより、結束する物品の継ぎ足しが容易となり、また、材料や建材などを結束し、運搬先で取り外し、何回でも再利用ができるものである。従って、完全な結束作業の迅速化、経済的効果を発揮する。

　本書では、結束される物品、電線、エアーホース、ロープ、チューブ、竹竿、サトウキビ、組み立てテントなどの結束の仕方をイラストで解説し、著者の説明を教授したテキスト教材でもある。

Preface

It is a tool which loosens the union band of the type whose reuse is impossible, and was made to be possible [reuse].

It can add and band together.

Therefore, speeding up of perfect union work and an economical effect are demonstrated.

They are also the text teaching materials which explained with the illustration the method of union of goods, an electric wire, an air hose, a rope, a tube, a bamboo pole, sugarcane, an assembly tent, etc. which band together in this book, and acted as a professor of an author's explanation.

目　次

１、多目的・結束バンド外し具（イラスト解説）

(1) 電線の結束 -- 5

(2) エアーホースの結束 -- 6

(3) ロープの結束 --- 7

(4) チューブの結束 -- 8

(5) 竹竿の結束 -- 9

(6) サトウキビの結束 ---10

(7) 組み立てテント材の結束 ----------------------------------11

２、英語解説

English of the usage

The removal instrument of a multiple-purpose union band

(illustration description)

(1) --12

Union of an electric wire

(2) --13

Union of an air hose

(3) --14

Union of a rope

(4) --15

Union of a tube

(5) --16

Union of a bamboo pole

(6) --17

Union of sugarcane

(7) ---18

Union of the pillar of a tent

3、中国語解説

多目的団結楽队的卸载器具(插图解说)

(1) ---19

电线的团结

(2) ---20

风管的团结

(3) ---21

绳的团结

(4) ---22

管的团结

(5) ---23

竹竿的团结

(6) ---24

甘蔗的团结

(7) ---25

装配帐篷木材的团结

4、公报解说---26

5、Patent journal English ---27

1、多目的・結束バンド外し具（イラスト解説）

⑴ 電線の結束

　細い電線、太い電線の結束に於いて、結束された束を継ぎ足す場合に取り外して、使用できる、継ぎ足しの束数の加減を調整する。

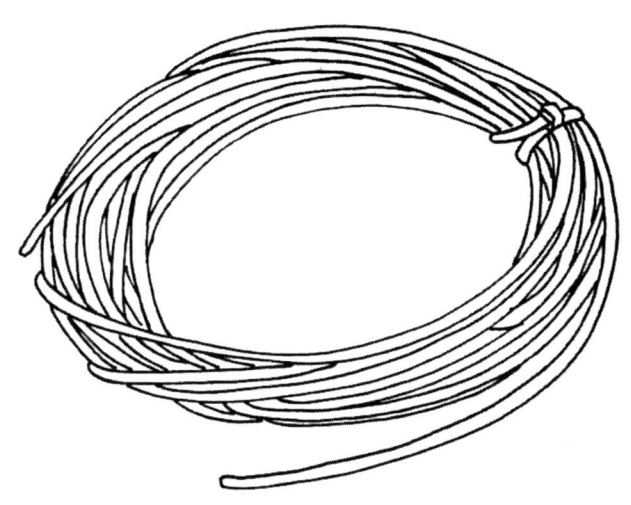

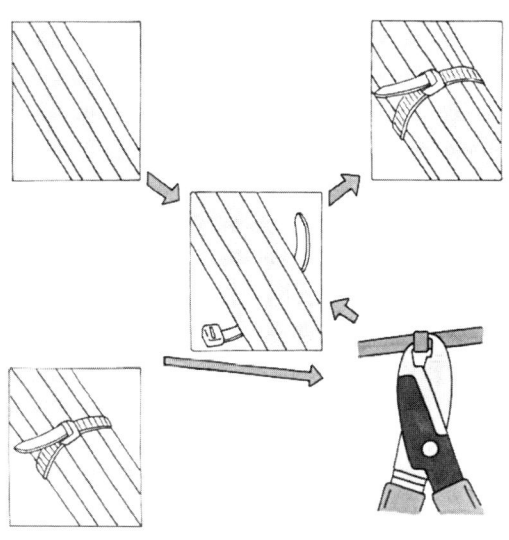

⑵ エアーホースの結束

　エアーホースの結束に老いて、エアーホースは電線と異なる結束バンドを使用する場合もあるので、継ぎ足しの束数の加減で調整する。

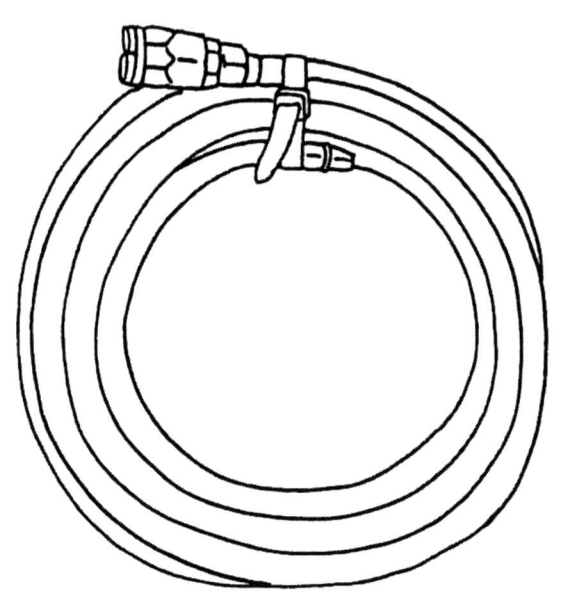

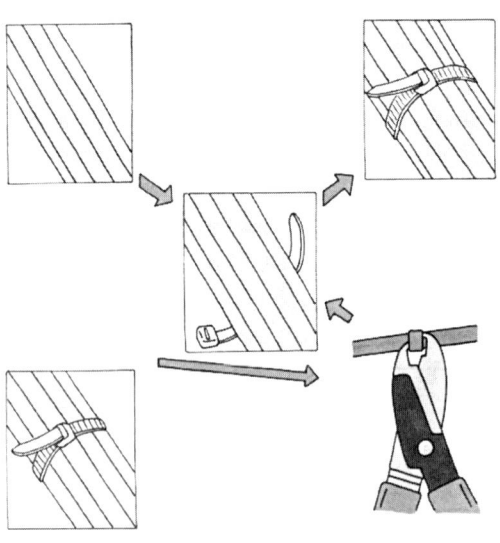

(3) ロープの結束

　ロープの結束の仕方は、電線やエアーホースと異なり、短いロープの結束で使用していた結束バンドを長いロープを使用したりする場合の取り外しに、この多目的・結束バンド外し具は重宝である。

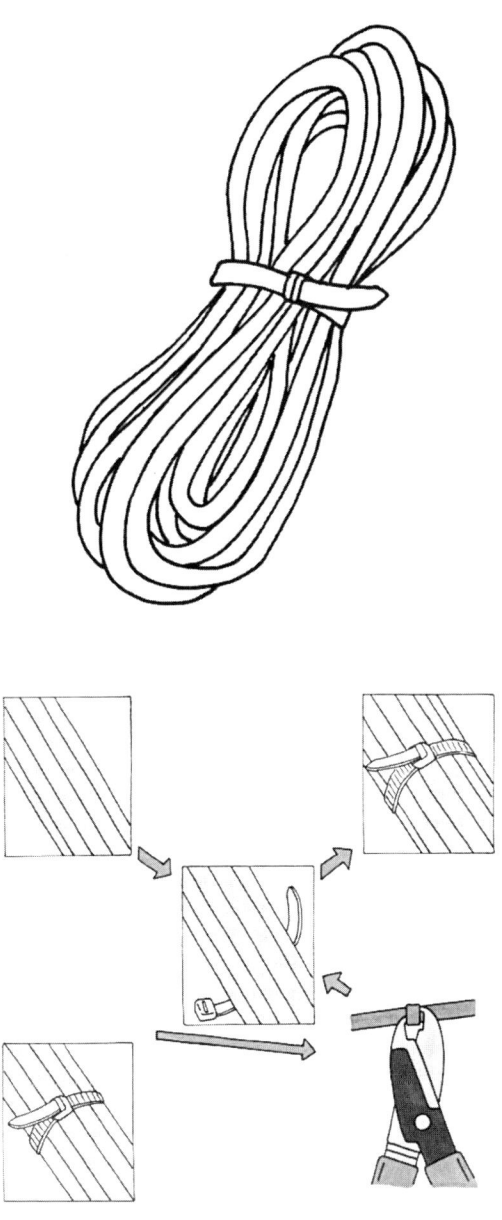

⑷ チューブの結束

車輪のチューブの結束は、比較的、頻繁に継ぎ足しを行う場合もあるので、多目的・結束バンド外し具の効果が期待できる。

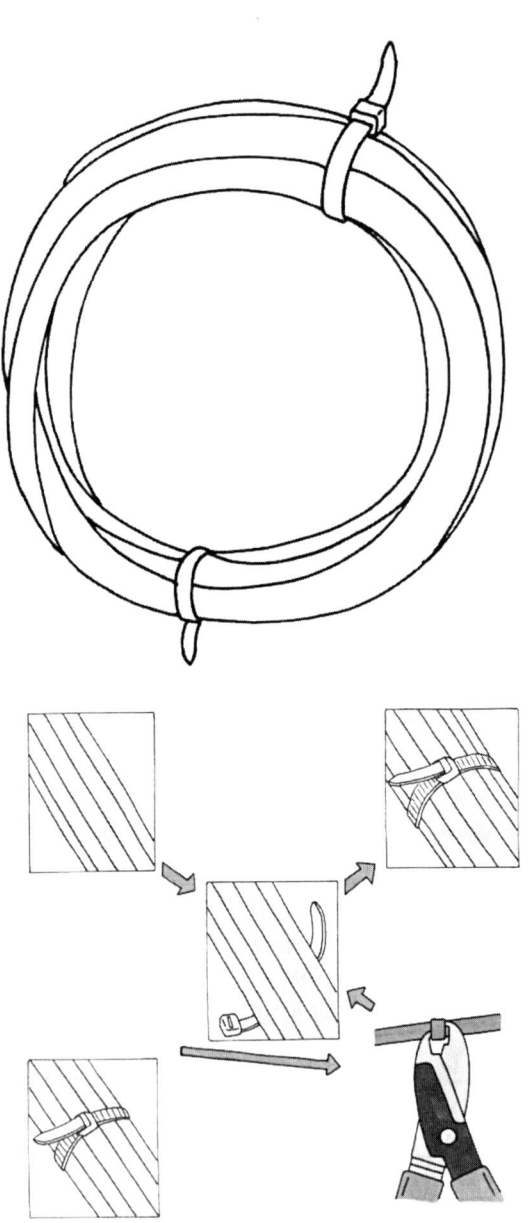

⑸ 竹竿の結束

　竹竿は結束箇所が2～3箇所になるので、継ぎ足しの場合、結束バンドの取り外しのスピード化が要求される。この多目的・結束バンド外し具はこれに対応したものであり、作業手順はイラストに示す。

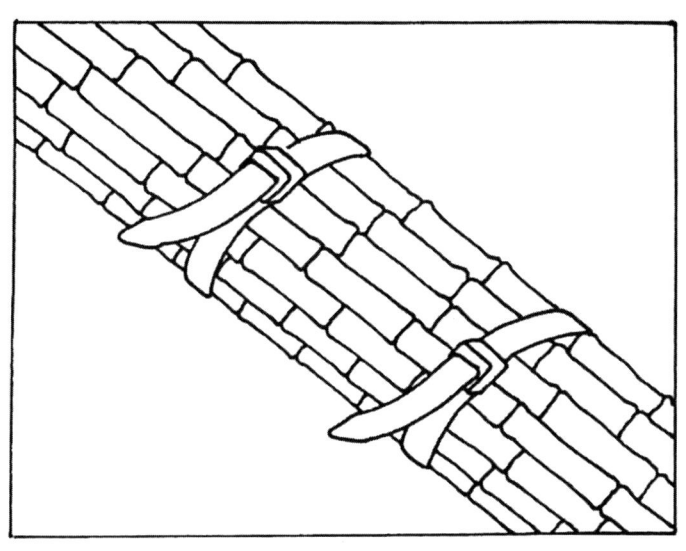

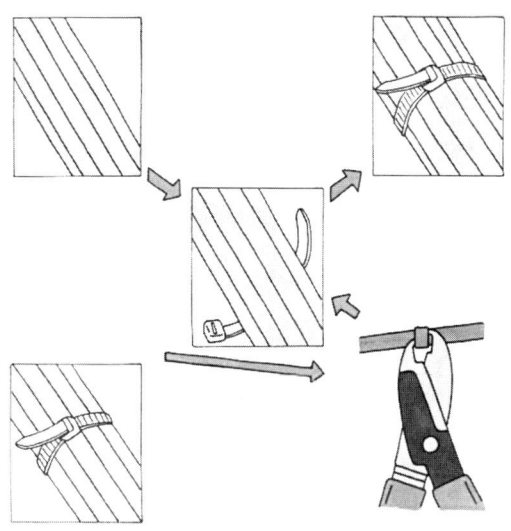

⑹ サトウキビの結束

　サトウキビの結束には、重宝である。重さで取引価格が決められる場合もあるので、減量、増量等の継ぎ足し作業に、この多目的・結束バンド外し具の効果が期待でき、納品後は再利用無しの使い捨て結束バンドとなる。

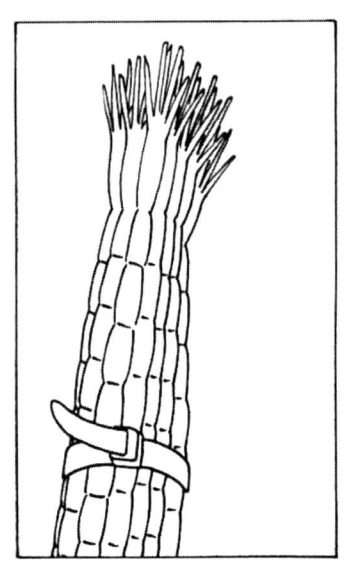

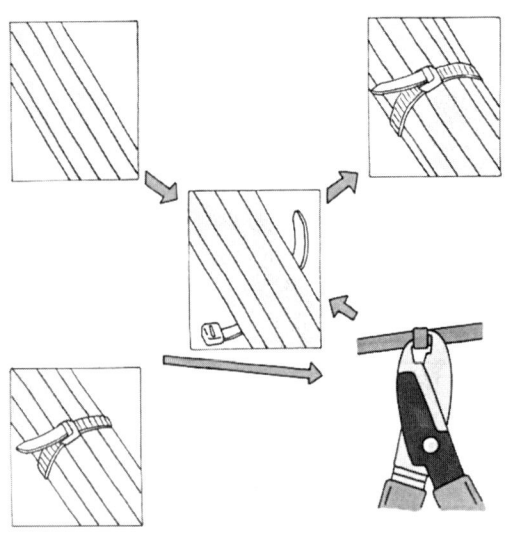

(7) 組み立てテント材の結束

　イベント会場などで、組み立てテントが多く使用されている。テント器材の柱パイプ、また竹材などの結束に安価な使い捨て結束バンドの再利用であれば、経済的効果が期待できる。

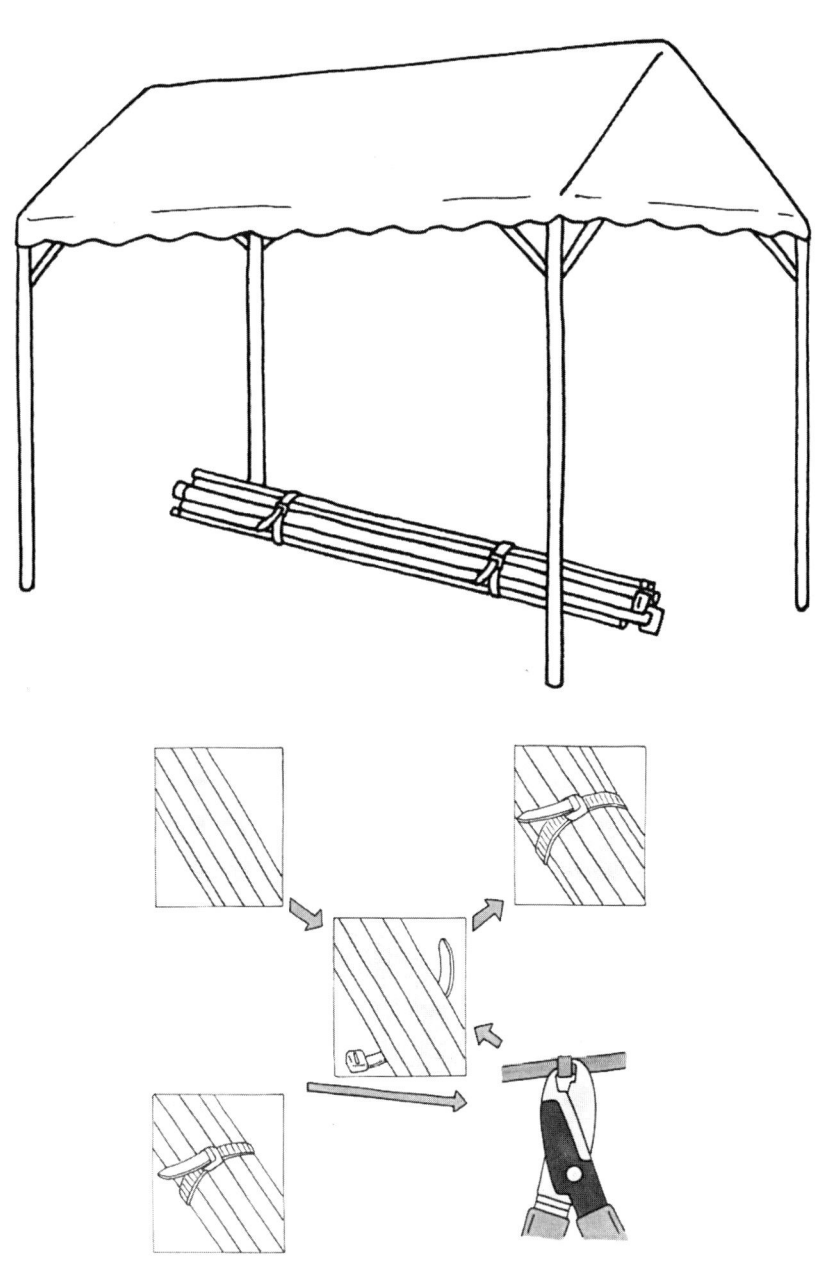

11

２、英語解説

English of the usage

The removal instrument of a multiple-purpose union band

(illustration description)

⑴

Union of an electric wire

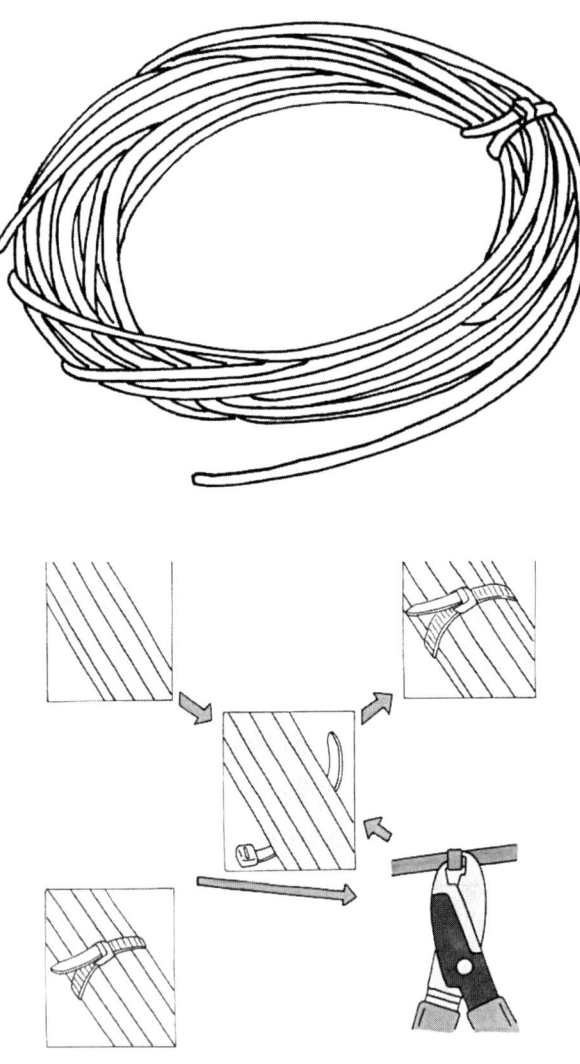

12

(2)

Union of an air hose

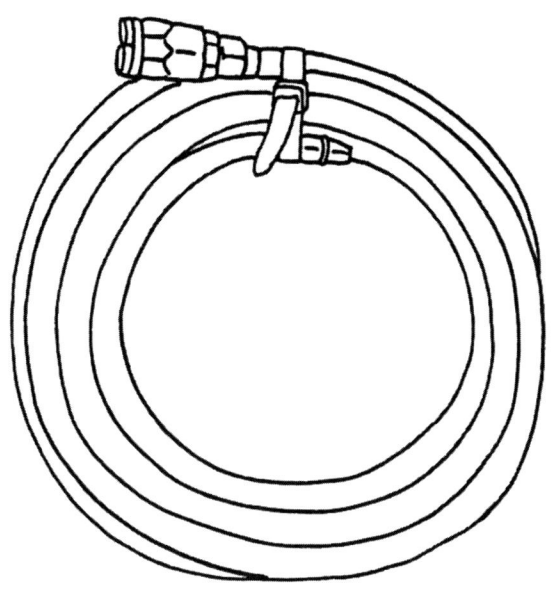

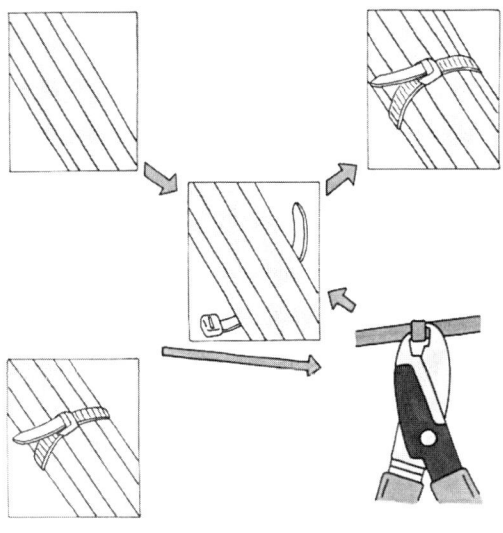

(3)

Union of a rope

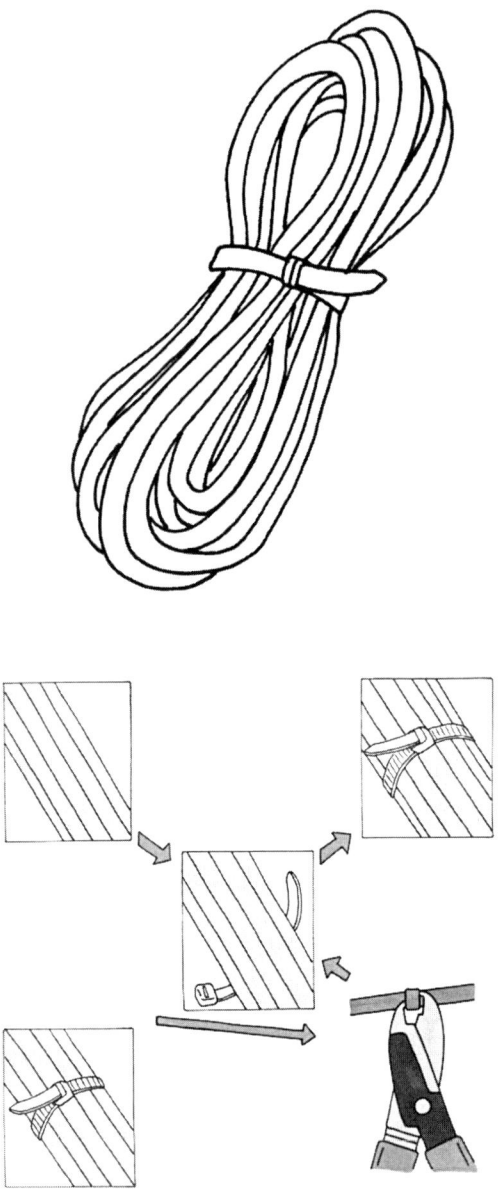

⑷ Union of a tube

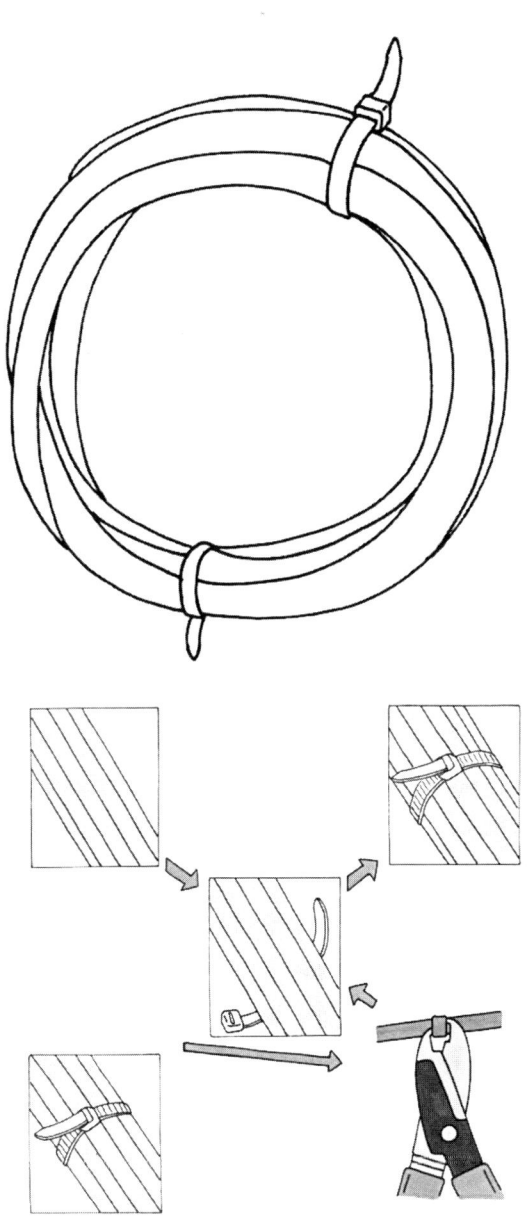

(5)

Union of a bamboo pole

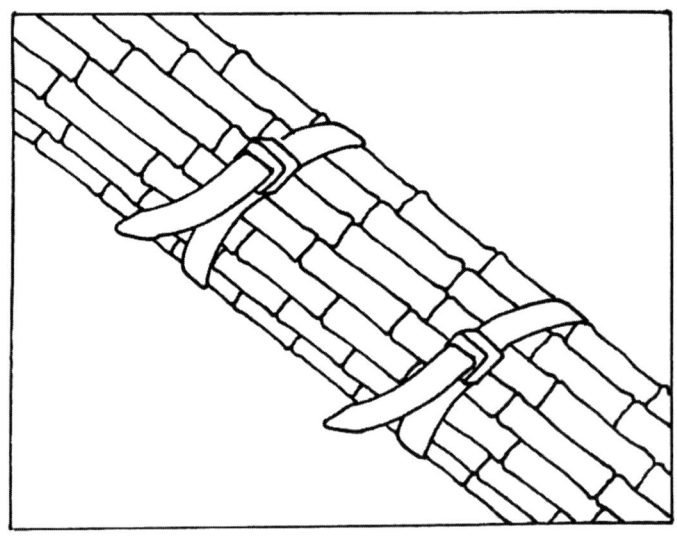

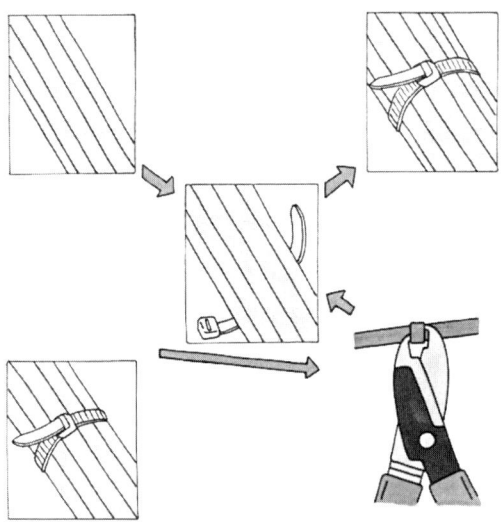

⑹

Union of sugarcane

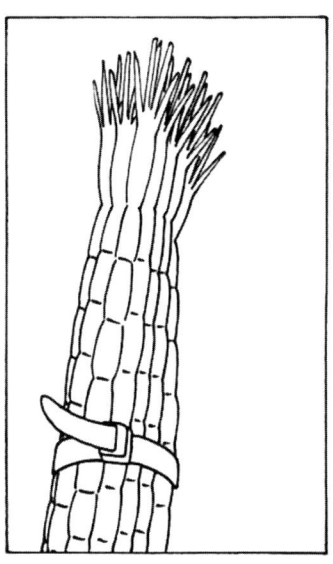

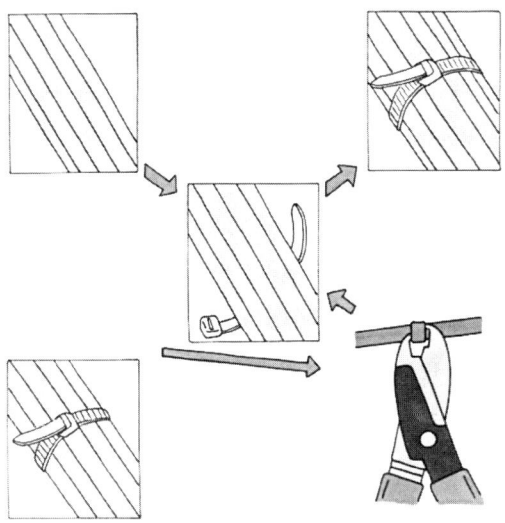

(7)

Union of the pillar of a tent

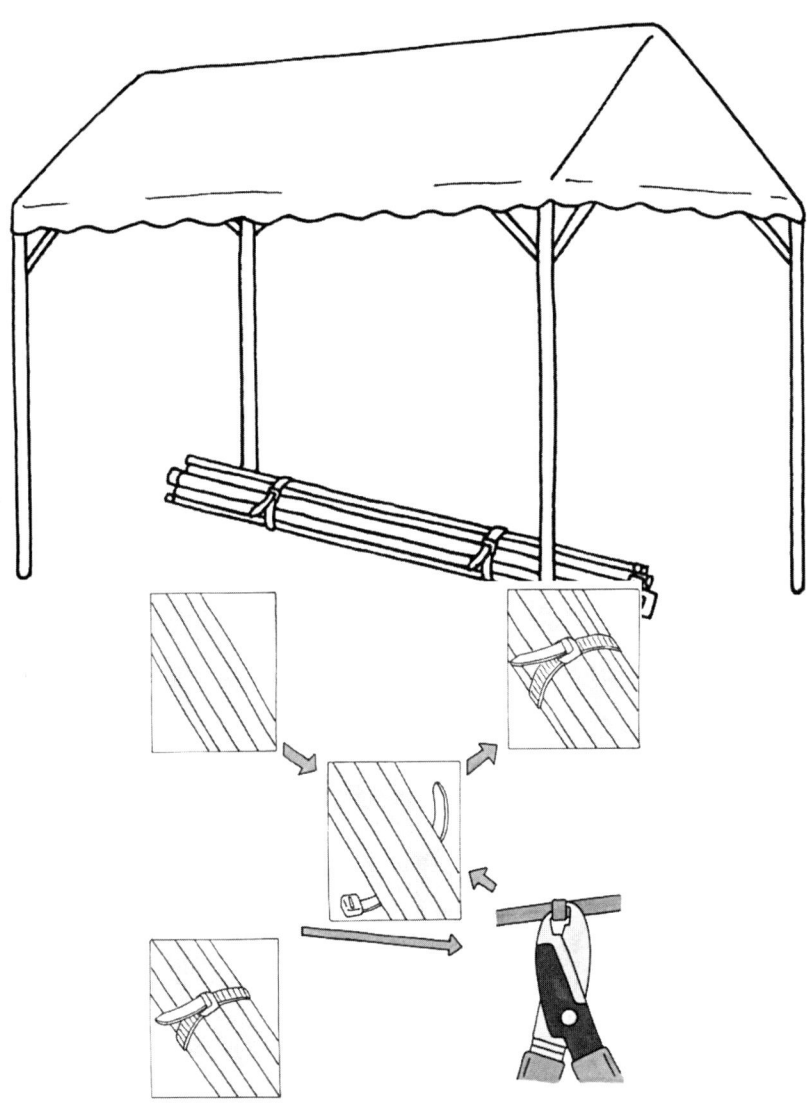

3、中国語解説

多目的团结乐队的卸载器具(插图解说)

⑴

电线的团结

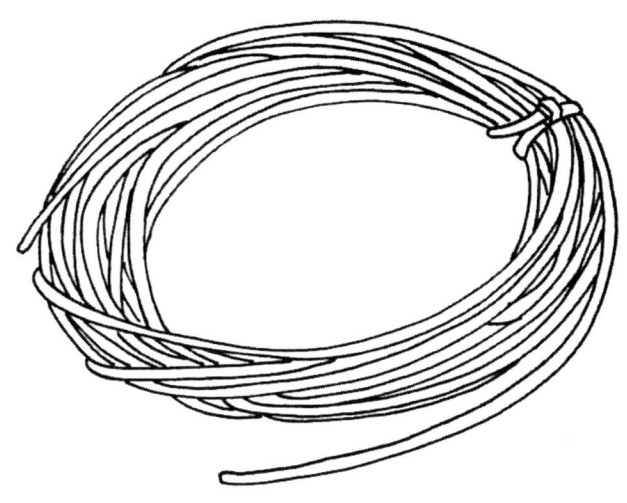

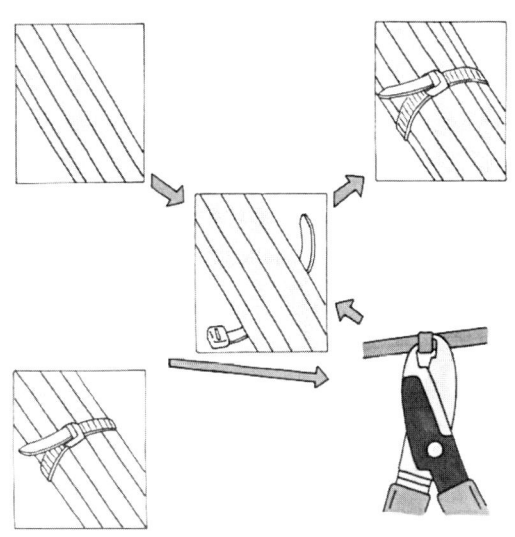

(2)

风管的团结

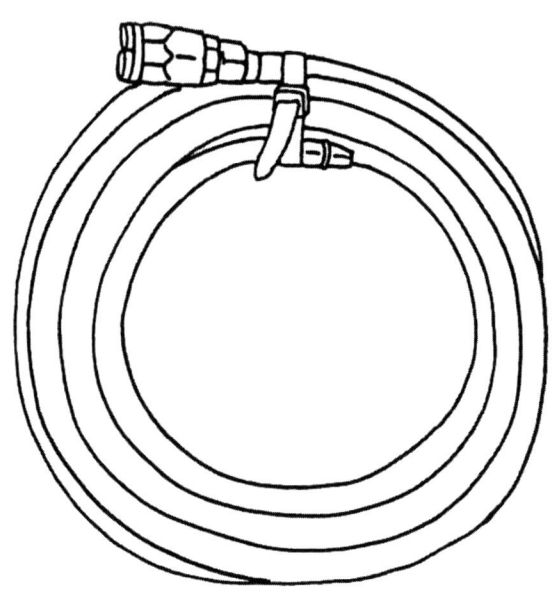

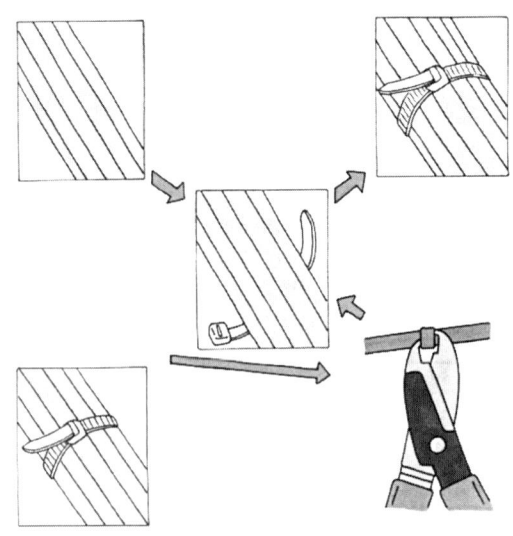

(3)

绳的团结

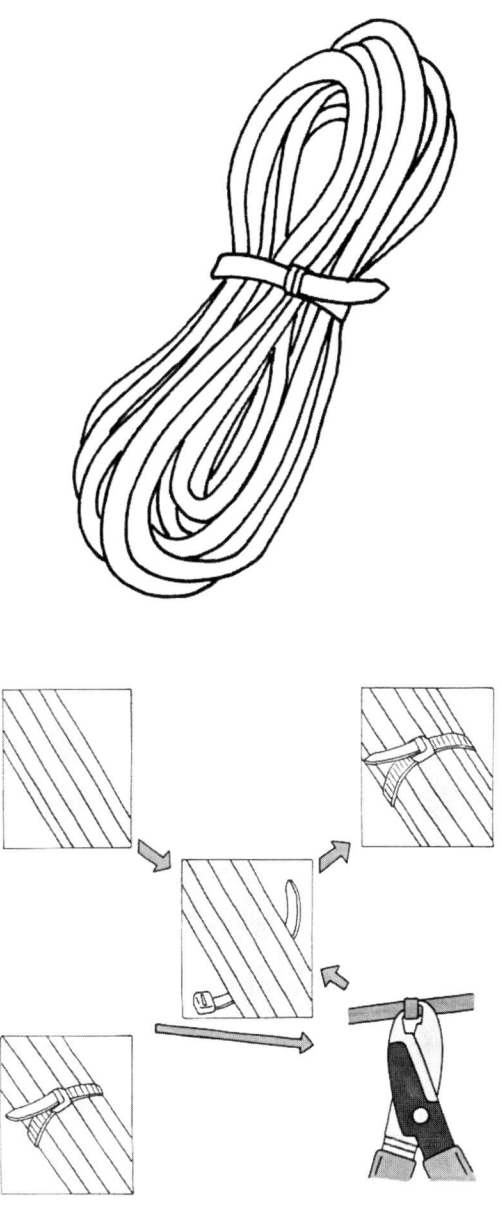

(4)

管的团结

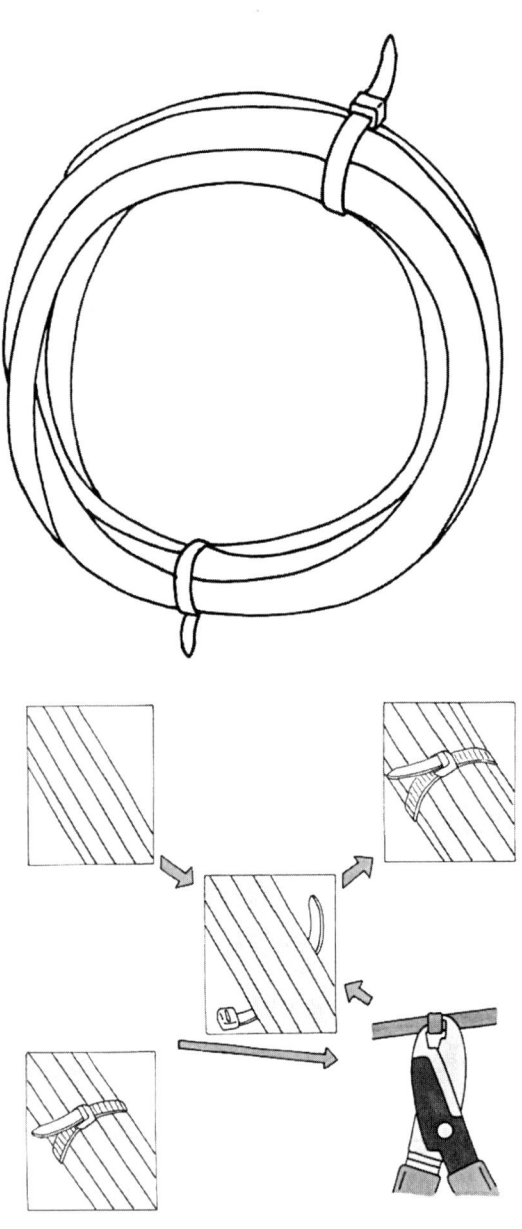

(5)

竹竿的团结

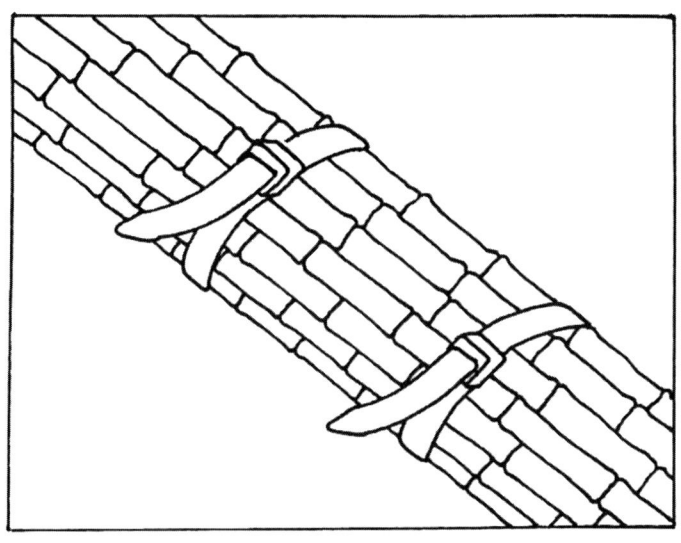

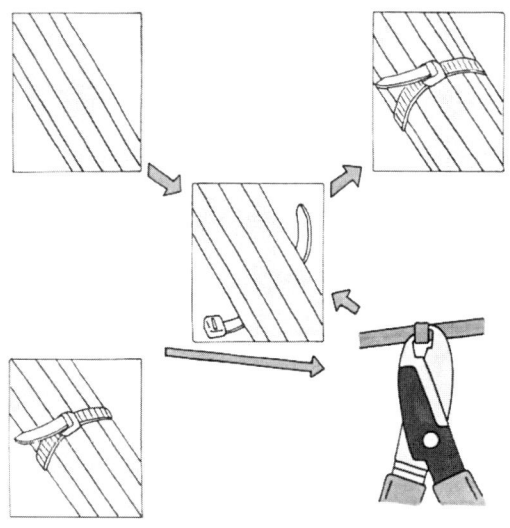

(6)

甘蔗的团结

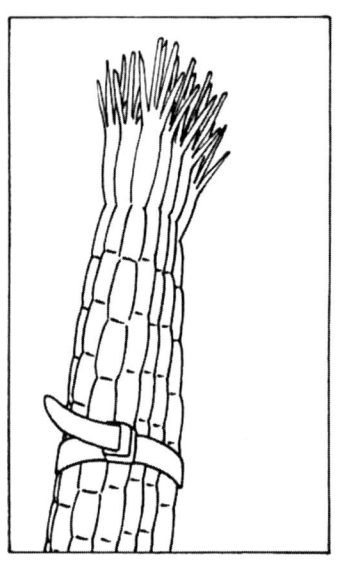

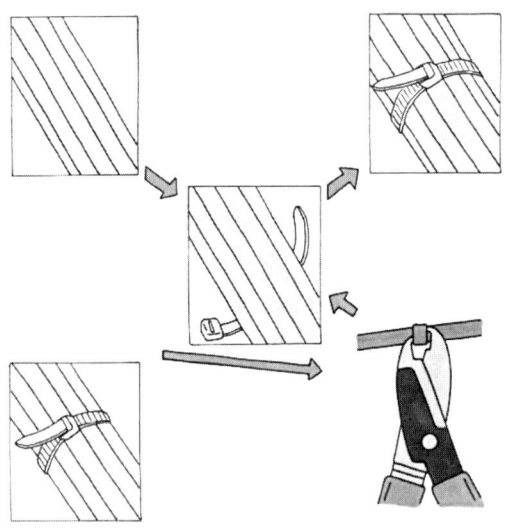

(7)

装配帐篷木材的团结

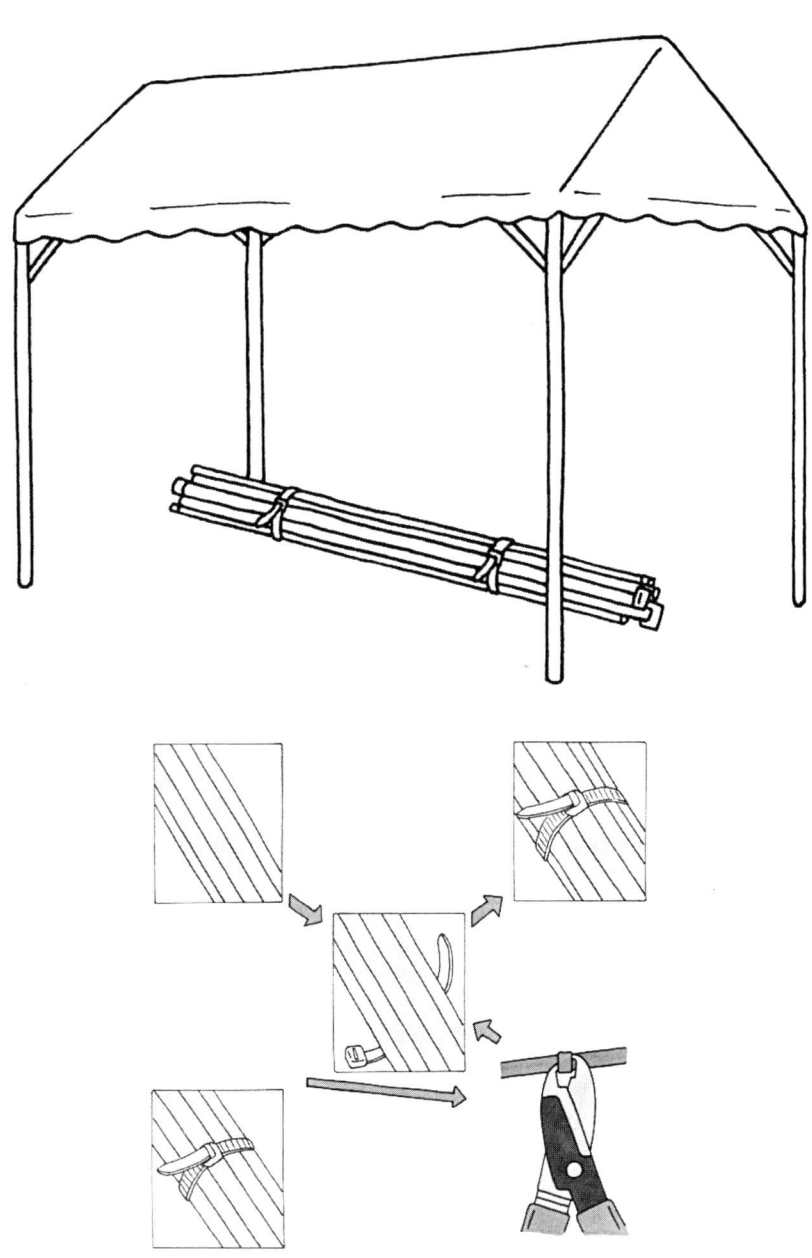

4、公報解説

実用新案登録第3183506号

考案の名称；結束バンド外し具

実用新案権者；富永 常夫

【要約】
【課題】安価かつ大量に出回っている一般的な結束バンド、すなわち一度締めると再利用できないタイプの結束バンドを緩めて取り外し、再利用可能とするための、結束バンド外し具を提供する。
【解決手段】一対のアーム13と、これらのアームが交差している交差部を回動自在に軸支する軸部14と、各アームの軸部から先端側に形成された一対の刃部11とを有し、アームを接近・離反すると、刃部が互いに接近・離反するようになった取り外し具において、一方の刃部の先端を細い棒状、且つ、他方の刃部に向けて屈曲した形状となし、他方の刃部の先端を細い棒状に形成した。
【選択図】図1

【実用新案登録請求の範囲】
【請求項1】
一対のアームと、これらのアームが交差している交差部を回動自在に軸支する軸部と、前記各アームの前記軸部から先端側に形成された一対の刃部とを有し、前記アームを接近・離反すると、前記刃部が互いに接近・離反するようになった取り外し具において、
一方の刃部（11）の先端部（11a）を細い棒状、且つ、他方の刃部に向けて屈

曲した形状となし、他方の刃部（１２）の先端部（１２ａ）を細い棒状に形成した結束バンド取り外し具。

【考案の詳細な説明】

【技術分野】

【０００１】

本考案は、コードや電線等を束ねて締めるときなどに使用する、いわゆる結束バンドの外し具に関するものである。

【背景技術】

【０００２】

従来から電気工事等においては、複数の電線などを結束するためのインシュロック（登録商標）と呼ばれる結束バンドが用いられている。この結束バンドは、係止ヘッド部とこれと一体に設けらたバンド部とからなり、このバンド部を被結束物の周囲に巻き付けると共に前記係止ヘッド部に形成されている空隙部に挿通することにより、この空隙部内に設けられたラチェット爪が前記バンド部に形成された凸凹に噛合して抜け止めされるものである。

【０００３】

上記従来の一般的な結束バンドは一度締めると緩めることが出来ず、再利用ができなかった。係る不都合を回避すべく、ヘッド部のカムを押すと、ロックがはずれ、取り外しが可能なものも存在するが、ヘッド部や帯の幅が相当広くなっていた。

【０００４】

特許文献１、２には再利用可能な結束バンドが開示されている。

【先行技術文献】

【特許文献】

【０００５】
【特許文献１】実開平７－１３７５６号公報
【特許文献２】特開２００９－９５２１８号公報
【考案の概要】
【考案が解決しようとする課題】
【０００６】
上記取り外し可能な結束バンドは、係止ヘッド部にカムを設けるなど、構造が複雑になりがちである。
【０００７】
本考案は、現在、安価かつ大量に出回っている一般的な結束バンド、すなわち一度締めると再利用できないタイプの結束バンドを緩めて取り外し、再利用可能とするための、結束バンド外し具を提供することを目的とする。
【課題を解決するための手段】
【０００８】
上記の目的を達成する本考案の構成は次の通りである。
【０００９】
一対のアームと、これらのアームが交差している交差部を回動自在に軸支する軸部と、前記各アームの前記軸部から先端側に形成された一対の刃部とを有し、前記アームを接近・離反すると、前記刃部が互いに接近・離反するようになった取り外し具において、一方の刃部の先端を細い棒状、且つ、他方の刃部に向けて屈曲した形状となし、他方の刃部の先端を細い棒状に形成した。
【考案の効果】
【００１０】

上記のように構成される本考案が、如何に作用して課題を解決するかを図面を参照しながら概説する。

【０　０　１　１】

図５は使用時における結束バンド１の係止ヘッド部２近傍の断面図である。係止ヘッド部２内にはラチェット爪２ａが設けられており、このラチェット爪２ａにバンド部３に設けられた凹凸３ａが噛合することでバンド部３が抜けない、すなわち図中左方向には移動しない構造となている。

【０　０　１　２】

図３、図４は、本考案の使用状態をあらわしている。なお、<u>図４</u>においては可視的にするため先端部１１ａ，１２ａ以外は省略してある。

【０　０　１　３】

結束バンド外し具１０の一方の刃部１２の先端部１２ａを係止ヘッド部２の側部にあてがった状態で、他方の刃部１１の屈曲された先端部１１ａを、係止ヘッド部２内に差込み、ラチェット爪２ａを押圧する。

【０　０　１　４】

すると、ラチェット爪２ａは折れ曲がり、バンド部３に設けられた凹凸３ａとの噛合が解放される。かかる状態であれば、バンド部３は係止ヘッド部２から抜ける方向、図中左方向に移動可能となり、バンド部３を係止ヘッド部２から抜くことで、当該結束バンド１の再利用が可能となるのである。

【図面の簡単な説明】

【０　０　１　５】

【図１】本考案の正面図

【図２】刃先を開いた状態の正面図

【図3】使用状態を示す説明平面図

【図4】使用状態を示す説明正面図

【図5】結束バンドのヘッド部の断面図

【考案を実施するための形態】

【0016】

以下、好ましい考案の一実施形態につき、図面を参照しながら概説する。なお、本考案の実施の形態は、下記の実施形態に何ら限定されることはなく、本考案の技術的範囲に属する限り種々の形態を採りうる。

【0017】

図1は本考案に係る結束バンド外し具10の閉じた状態の正面図であり、図2は開いた状態の正面図である。

【0018】

これらの図に示すように、本考案の結束バンド外し具10は、一対のアーム13,13が軸部14で交差して回動自在に連結され、このアーム13,13の軸部14から先端側に、刃部11,12を形成し、各刃部11、12の先端に先端部11a,12aを形成している。アーム13,13の後端側はカバーに覆われたハンドル15,15となっている。ハンドル15,15を接近・離反すると、先端部11a,12aも互いに接近・離反するようになっている。

【0019】

本考案の特徴は先端部11a,12aの形状にある。係止ヘッド部2に嵌入される先端部11aは、他方の刃部12側に向かって屈曲された、且つ、細い棒状を呈している。係止ヘッド部2の側壁に当接される先端部12aは細い棒状を呈している。

【符号の説明】

【0020】

10・・結束バンド外し具　11・・刃部　11a・・先端部

12・・刃部　12a・・先端部　13・・アーム

14・・軸部　15・・ハンドル

【図面の簡単な説明】

【0015】

【図1】本考案の正面図

【図2】刃先を開いた状態の正面図

【図3】使用状態を示す説明平面図

【図4】使用状態を示す説明正面図

【図5】結束バンドのヘッド部の断面図

【図1】

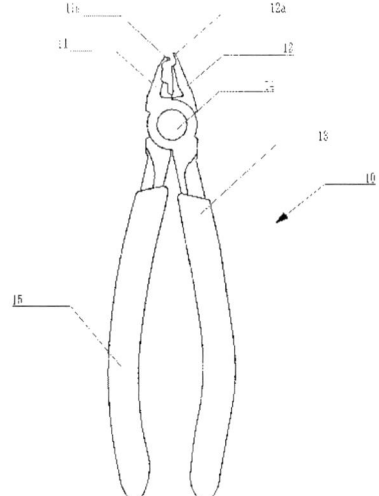

【図2】

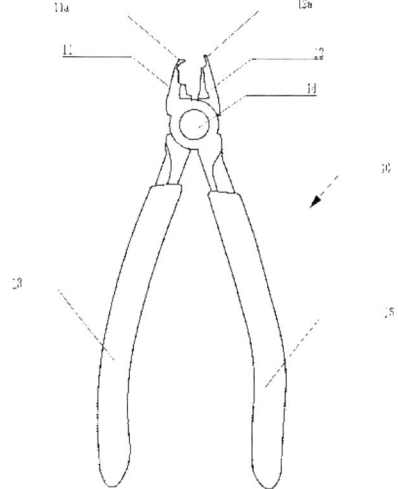

【図3】

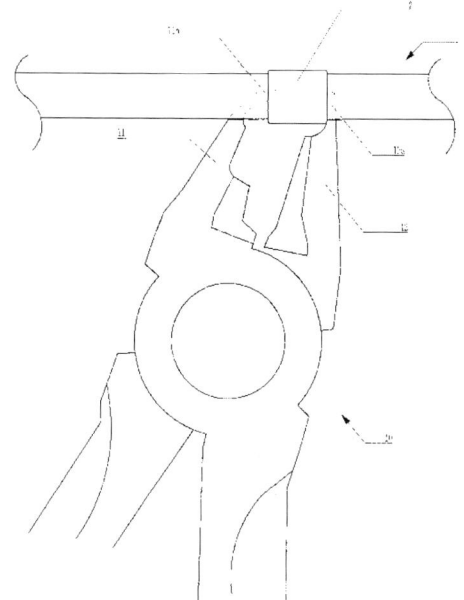

【図4】

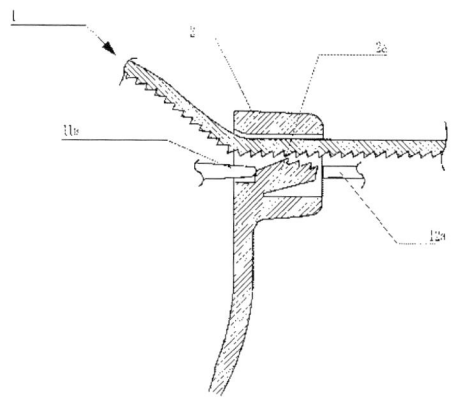

【図5】

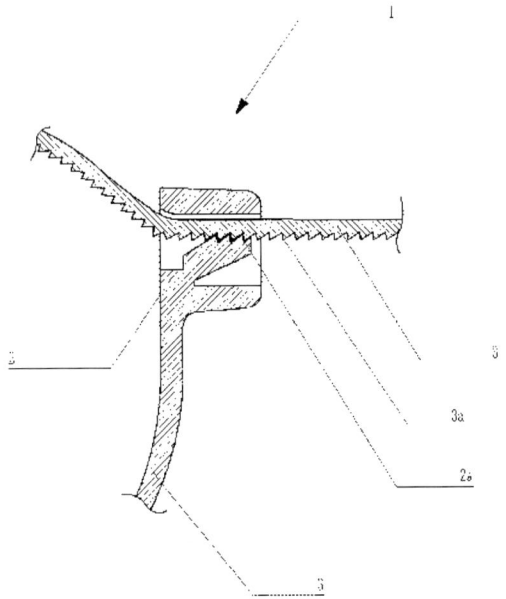

5. Patent journal English

[Claims]

[Claim 1]

In a removal implement with which the aforementioned cutting part came to approach and separate mutually when it has a pair of arm, a shank which supports pivotally an intersection which these arms intersect enabling free rotation, and the pair of cutting part formed in the tip side from the aforementioned shank of each aforementioned arm and the aforementioned arm was approached and separated,

It is the thin binding band removal implement formed cylindrically about a point (12a) of a cutting part (12) of form which turned a point (11a) of one cutting part (11) to thin rod form and a cutting part of another side, and was bent, nothing, and another side.

DETAILED DESCRIPTION

[Detailed explanation of the device]

[Field of the Invention]

[0001]

What is called a binding band used when bundling and fastening a code, an electric wire, etc. removes this design, and it is related in detail.

[Background of the Invention]

[0002]

The binding band called the insulation lock (registered trademark) for banding two or more electric wires together in an electrical work etc. from the former is used. By inserting in the cavity part which provide this binding band integrally with a locking head part and this, it consists of a **** band part, and this band part is twisted around the circumference of a thing to be banded together, and is formed in the aforementioned locking head part, The ratchet claw provided in this cavity part is the thing which was formed in the aforementioned band part and which gears unevenly and is locked.

[0003]

Once it fastened the above-mentioned conventional common binding band, it could not be loosened, and reuse of it was not completed. That the starting inconvenience should be avoided, when the cam of the head section was pushed, the lock separated, and what can be removed existed, but the width of a head section or a belt was fairly wide.

[0004]

The recyclable binding band is disclosed in a Patent document 1 and 2.

[Citation list]

[Patent literature]

[0005]

[Patent document 1] JP,H7-13756,U

[Patent document 2] JP,2009-95218,A

[The outline of a device]

[Problem(s) to be Solved by the Device]

[0006]

the above -- structures -- a dismountable binding band provides a cam in a locking head part -- tend to become complicated -- it comes out.

[0007]

This design loosens and removes the binding band of the type which is not recyclable once the common binding band now which has appeared on the market inexpensive and in large quantities, i.e., fasten, and aims at the binding band for making reuse possible outside's carrying out, and providing an ingredient.

[Means for solving problem]

[0008]

The composition of this design which attains the above-mentioned purpose is as follows.

[0009]

A pair of arm and the shank which supports pivotally the intersection which these arms intersect enabling free rotation, In the removal implement with which the aforementioned cutting part came to approach and separate mutually when it has the pair of cutting part formed in the tip side from the aforementioned shank of each aforementioned arm and the aforementioned arm was approached and separated, The tip of the cutting part of the form which turned the tip of one cutting part to thin rod form and the cutting part of another side, and was bent, nothing, and another side was formed in thin rod form.

[Effect of the Device]

[0010]

It is outlined how this design constituted as mentioned above acts, and solves problem, referring to Drawings.

[0011]

Fig.5 is an about two locking head part [of the binding band 1 at the time of use] cross sectional view. Structure and ******** which the ratchet claw 2a is provided in the locking head part 2, and the band part 3 does not fall out because the unevenness 3a provided by this ratchet claw 2a at the band part 3 gears, i.e., do not move in the direction of the left in the figure.

[0012]

Fig.3 and Fig.4 express the busy condition of this design.In order to make it in visible in Fig.4, it has omitted except the point 11a and 12a.

[0013]

Where binding band outside carried out and the point 12a of one cutting part 12 of the ingredient 10 is assigned to the side part of the locking head part 2, the point 11a in which the cutting part 11 of another side was bent is inserted in the locking head part 2, and the ratchet claw 2a is pressed.

[0014]

Then, the ratchet claw 2a bends and engagement with the unevenness 3a provided by the band part 3 is released. If it is in this state, the band part 3 becomes movable in the direction which escapes from the locking head part 2, and the direction of the left in the figure, it will be extracting the band part 3 from the locking head part 2, and the reuse of the binding band 1 concerned of it will be attained.

[Brief Description of the Drawings]

[0015]

[Drawing 1]The front view of this design

[Drawing 2]The front view in the state where the edge of a blade was opened

[Drawing 3]The description plan view showing a busy condition

[Drawing 4]The description front view showing a busy condition

[Drawing 5]The cross sectional view of the head section of a binding band

[The form for devising]

[0016]

Hereafter, it gives an outline, referring to the Drawings per one embodiment of a preferable device. The embodiment of this design is not limited to the following embodiment at all, and as long as it belongs to technical scope of this design, it can take various forms.

[0017]

Fig.1 is a front view in the state where the binding band concerning this design outside carried out, and the ingredient 10 closed, and Fig.2 is a front view in the state where it opened.

[0018]

As shown in these figures, the binding band of this design outside carries out, and the pair of arms 13 and 13 cross by the shank 14, and the ingredient 10 is connected, enabling free rotation, forms the cutting parts 11 and 12 in the tip side from the shank 14 of these arms 13 and 13, and forms the points 11a and 12a at the tip of each cutting part 11 and 12. The back end side of the arms 13 and 13 serves as the handles 15 and 15 covered with covering. If the handles 15 and 15 are approached and separated, the points 11a and 12a will also approach and separate mutually.

[0019]

The characteristics of this design are in the form of the points 11a and 12a. The point 11a inserted in the locking head part 2 is bending and presenting thin rod form toward the cutting part 12 side of another side. The point 12a by which the side wall of the locking head part 2 is abutted is presenting thin rod form.

[Explanations of letters or numerals]

[0020]

10 .. Binding band outside carries out and it is ingredient 11.. Cutting part 11a .. Point

12 .. Cutting part 12a .. Point 13 .. Arm

14 .. Shank 15 .. Handle

[Drawing 1]

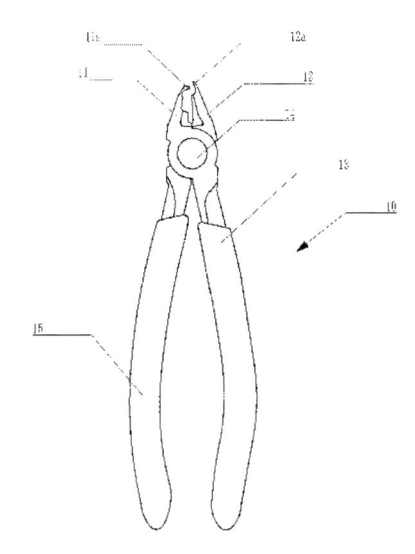

[Drawing 2]

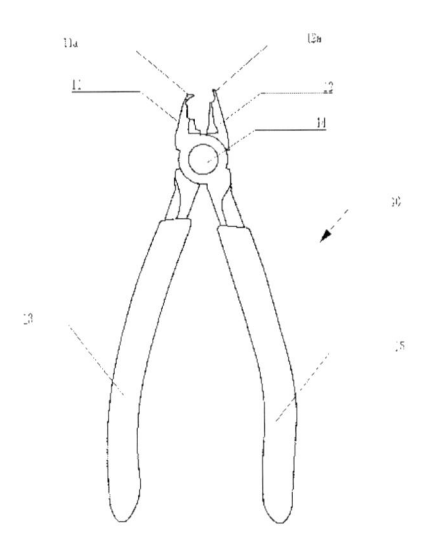

[Drawing 3]

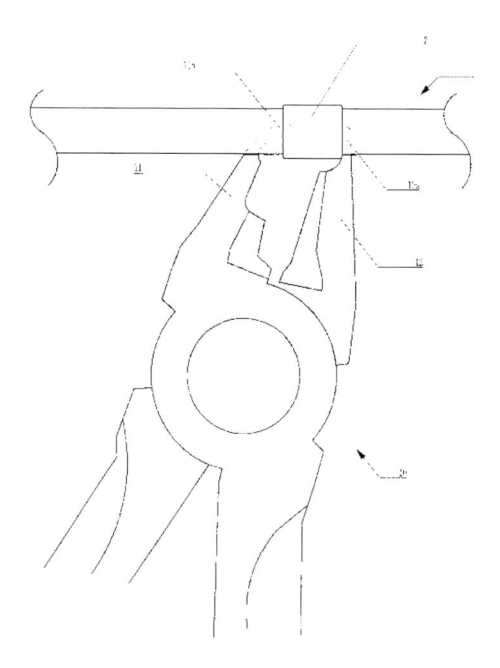

[Drawing 4]

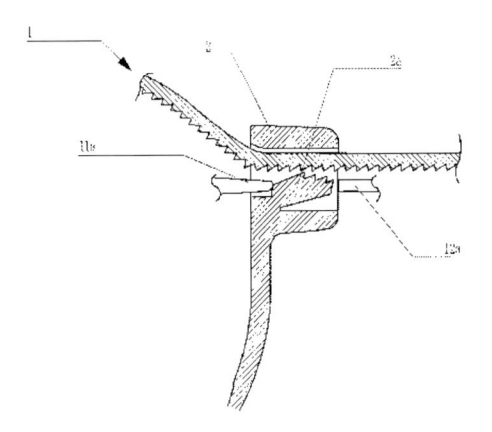

[Drawing 5]

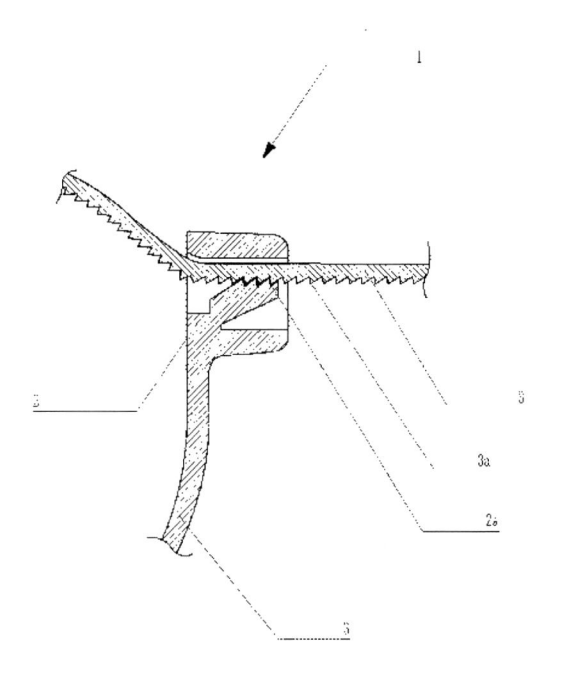

あとがき

　この多目的・結束バンド外し具は、私が長年、電線やエアーホースを束ねる作業に於いて、常時使用していましたが、追加で縛る或いは、取り付け変更や仮止めなどの時には、切断して捨てるしかありませんでした。その時「もったいない」と思うことが時々あり、何とか外してもう一度締め直しが出来ないか？と思い。今回の取り外し具の開発に至りました。当初は薄い銅板などギザギザな所とリップ部分に挟みロックが効かないようにして外してみましたが、手間が掛かり現実的ではないと思いました。次にたまたま錆びた使えないニッパーが有ったので、そのニッパーの先端部を削って（現在の形）挟んで見たら、上手く行きました。

　結束バンドも今では世界で使われるようになってきましたが、この工具は手で外せないタイプの結束バンドをも簡単に取り外しが可能で再利用させるための道具です。但し、思いっきり縛った時（バンドのラチェット部分でのストレスの変形）や余分な所を切断したバンド（挿入しにくい）或いは、購入後かなり日数が経って材質の劣化している等は再利用に不向きです。

　使用方法は輪になったバンドの四角い頭部分のラチェットを中と外から軽く外し具にて挟むだけでラチェット部分の解除ができ、輪を広げるように引っ張ると外せます。

　前記の商品等の増減の結束において、一度外した結束バンドのラチェット部が元の位置に戻らずロック不良が発生することがありますが、これらは、外し具を使用しているとバンドを締める復元方法がわかります。

　経歴、昭和 46 年、日産系ディーラーサービス業務 10 年勤務、工作機械仕上げ技術開発の経験のち現在の専用機械組み立て技術に従事。その他、20 歳で手作りホバークラフト制作、水陸両用車（人力）制作、人力＆ソーラーボート設計制作、レース出場にトライ。

著者　伊吹哲太郎（いぶきてつたろう）

多目的・結束バンド外し具の用途解説

定価（本体 1,500 円＋税）

２０１４年（平成２６年）４月15日発行

No. ＩＢＴ-019

発行所　IDF（INVENTION DEVLOPMENT FEDERATION）
　　　　発明開発連合会®
メール　03-3498@idf-0751.com　www.idf-0751.com
電話　03-3498-0751㈹
150-8691 渋谷郵便局私書箱第２５８号
発行人　ましば寿一
著作権企画　IDF 発明開発(連)
Printed in Japan
著者　伊吹哲太郎 ©

本書の一部または全部を無断で複写、複製、転載、データーファイル化することを禁じています。

It forbids a copy, a duplicate, reproduction, and forming a data file for some or all of this book without notice.